FLAMINGOS
AT THE ZOO

written and photographed
by Mia Coulton

"Honk, honk, honk!"

The flamingos are making a lot of noise at the zoo.

3

Flamingos are birds that are covered with pink, orange or red feathers.

They get their color from the foods they eat.

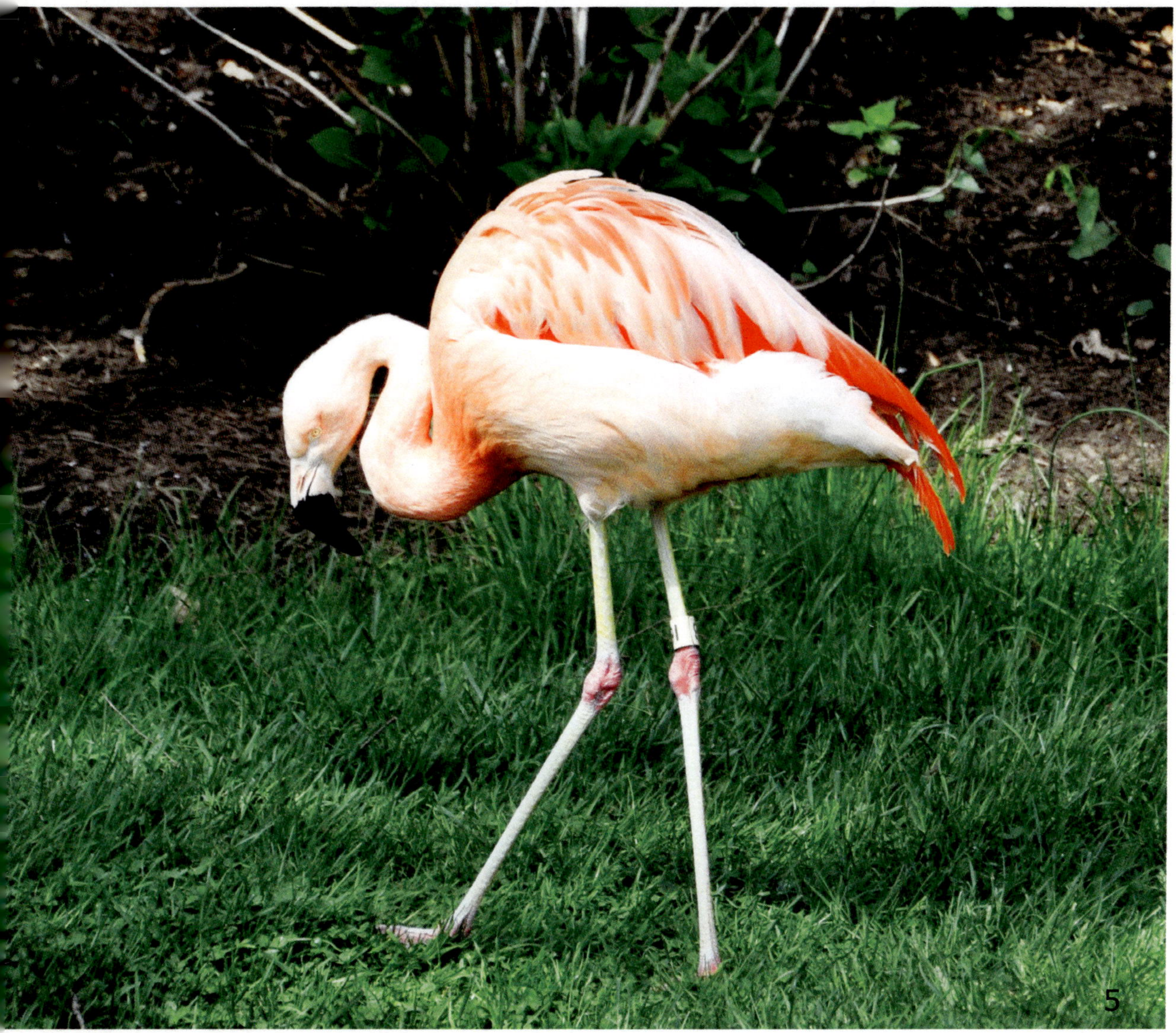

Flamingos have long, S-shaped **necks** and long legs like **stilts**.

The legs of a flamingo are longer than the flamingo's body.

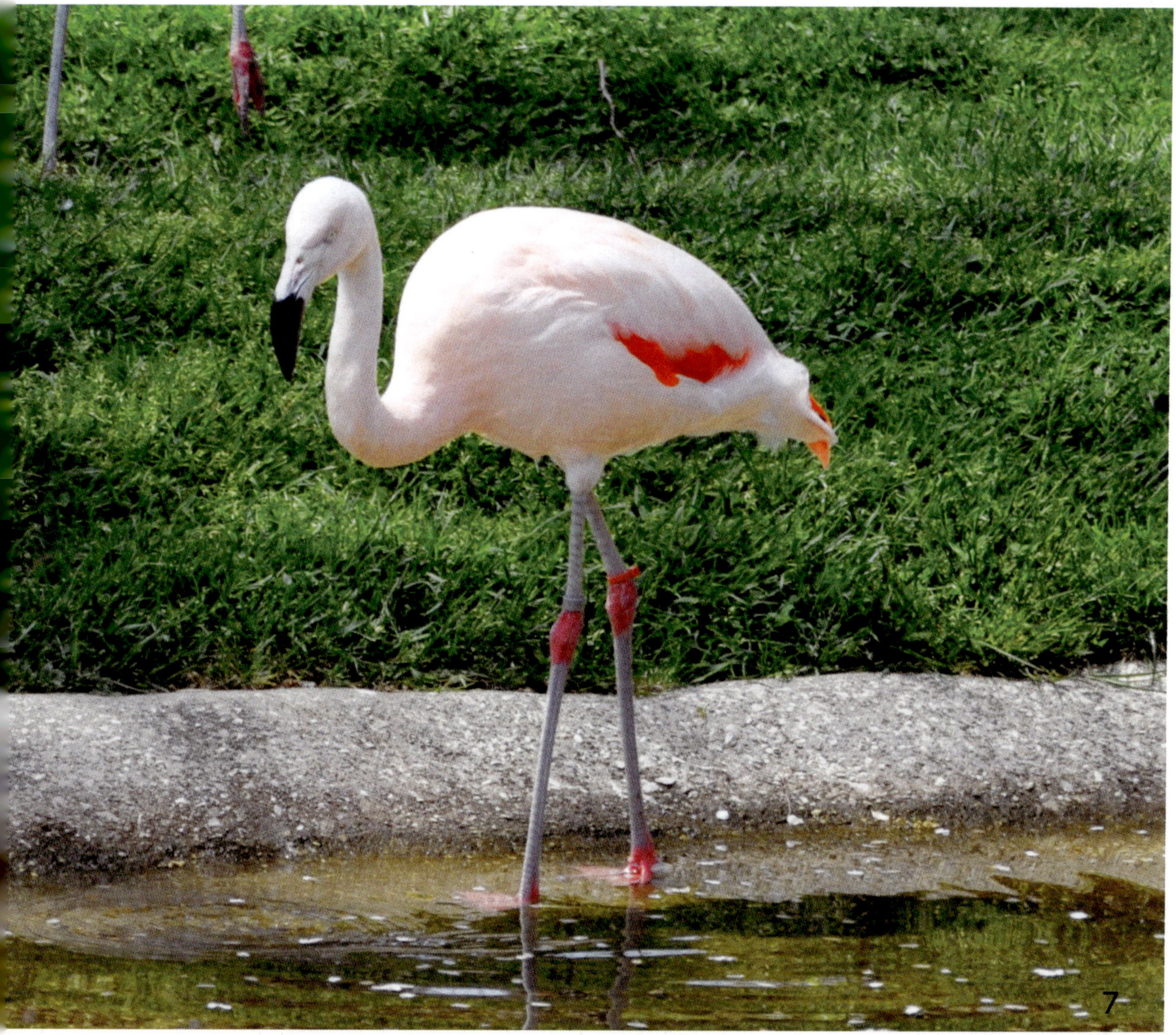

A flamingo can stand on one leg for a long time.

The other leg is tucked up and under its **wing**.

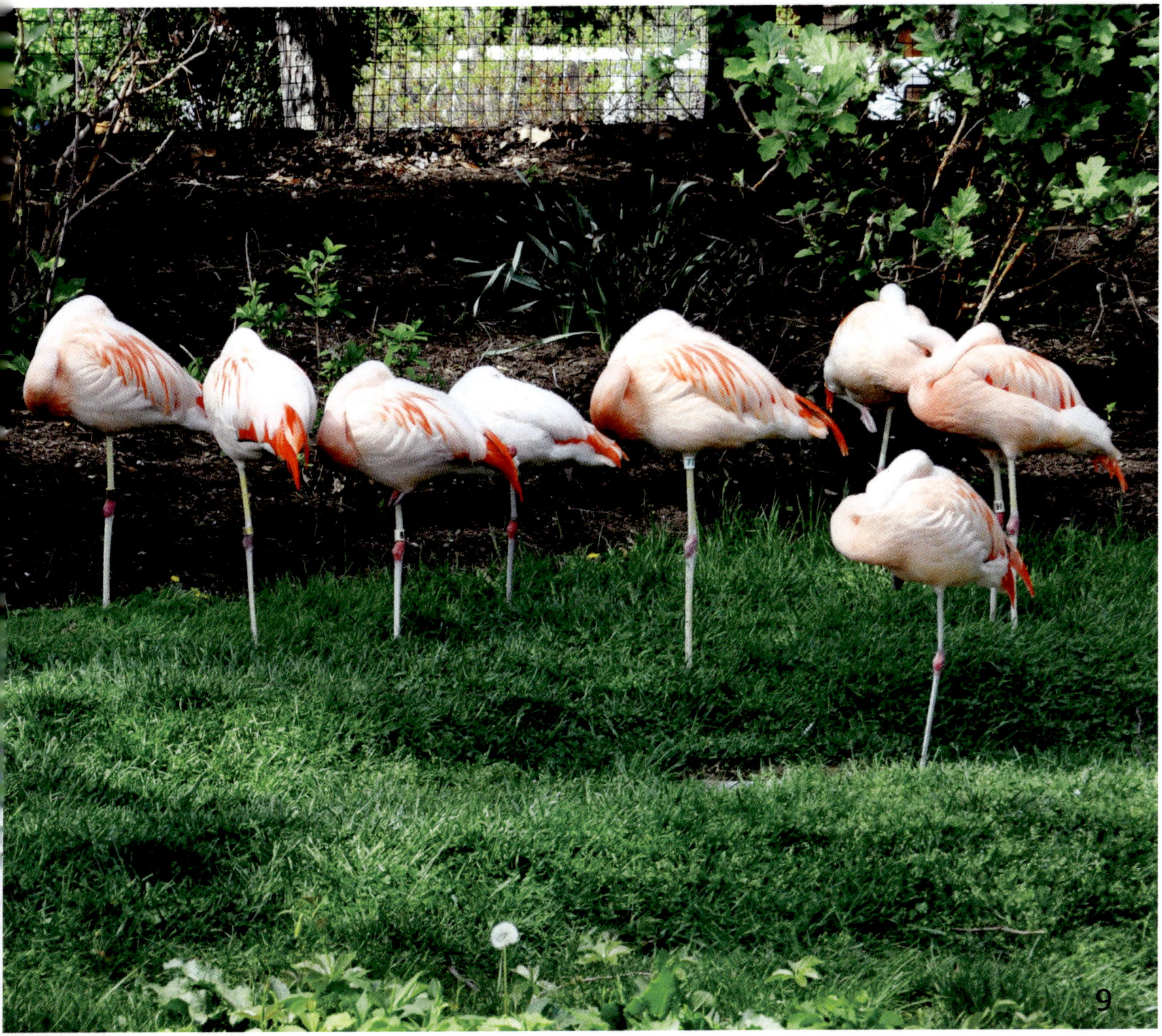

The flamingo has a **curved** beak that makes drinking water and scooping up food easy for the flamingo.

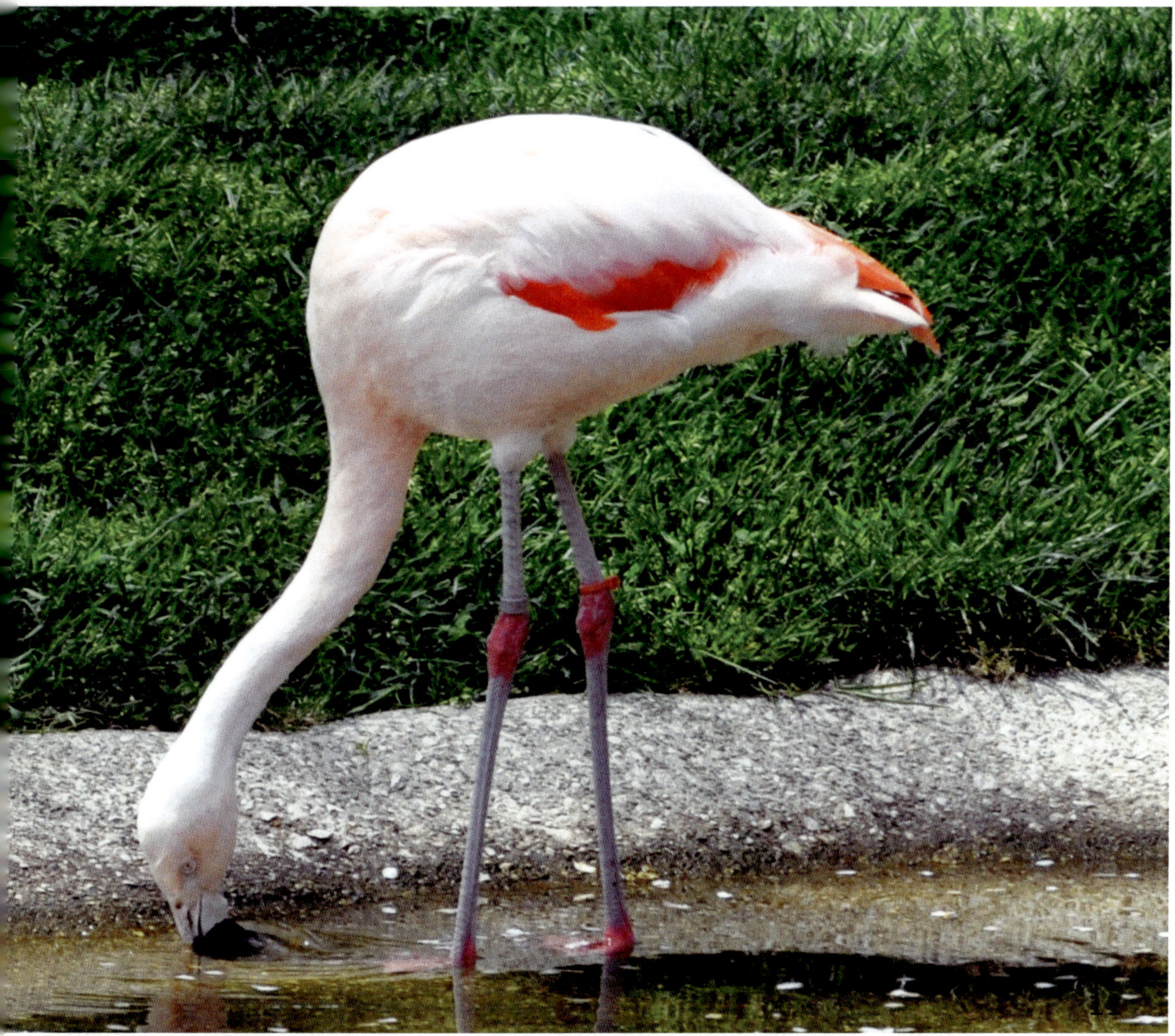

Flamingos have two wings and can fly.

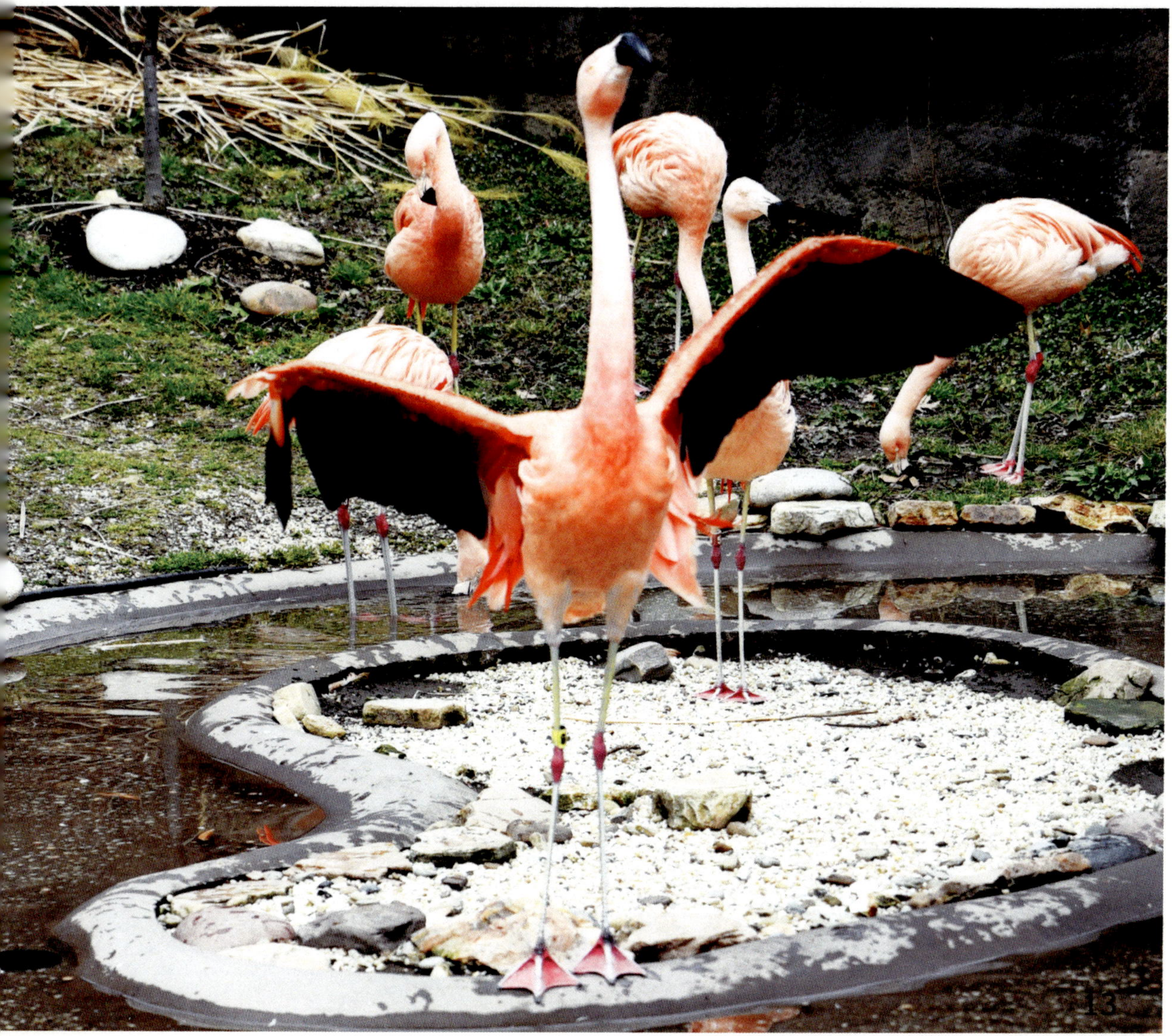

Flamingos do not like to be alone at the zoo.

Flamingos like to be with other flamingos at the zoo.

15

Glossary

curved: a shape with a long, smooth bend

feathers: the lightweight growths that cover the bodies of birds

necks: more than one neck, the part of the body that connects the head to the rest of the body

stilts: thin posts that hold something above the surface of land or water

wing: the feathered body part of a bird, used for flying and gliding